SPOTLIGHT ON GLOBAL ISSUES

EARTH'S PLASTIC PROBLEM

Adam Furgang

ROSEN PUBLISHING
NEW YORK

Published in 2022 by The Rosen Publishing Group, Inc.
29 East 21st Street, New York, NY 10010

First Edition

Editor: Theresa Emminizer
Book Design: Michael Flynn

Photo Credits: Cover chaiyapruek youprasert/Shutterstock.com; (series globe background) photastic/Shutterstock.com; p. 5 Maxim Blinkov/Shutterstock.com; p. 6 Apic/Hulton Archive/Getty Images; p. 7 Jeff Greenberg/Universal Images Group/Getty Images; p. Ekaterina43/Shutterstock.com; p. 9 https://en.wikipedia.org/wiki/Nylon_riots#/media/File:Standing_in_line_for_nylon_stockings_at_Millers_dept._store_Oak_Ridge_(7851145488).jpg; p. 11 (background) Alina G/Shutterstock.com; pp. 11, 13 (poster) courtesy of Library of Congress; p. 12 https://en.wikipedia.org/wiki/File:Richard_Nixon_presidential_portrait.jpg; p. 15 Jag_cz/Shutterstock.com; p. 16 Jekurantodistaja/Shutterstock.com; p. 17 Mohamed Abdulraheem/Shutterstock.com; p. 19 Bloomberg/Getty Images; p. 21 Allen J. Schaben/Los Angeles Times/Getty Images; p. 22 Justyna Sobesto/Shutterstock.com; p. 23 Andreas Rentz/Getty Images; p. 24 Michel Porro/Getty Images; p. 25 Robin Utrecht/AFP/Getty Images; p. 27 (both) Charlie Crowhurst/Getty Images; p. 29 Visharo/Shutterstock.com; p. 30 Tuomas Lehtinen/Shutterstock.com.

Library of Congress Cataloging-in-Publication Data

Names: Furgang, Adam, author.
Title: Earth's plastic problem / Adam Furgang.
Description: New York : Rosen Publishing, [2022] | Series: Spotlight on global issues | Includes index.
Identifiers: LCCN 2019052501 | ISBN 9781725323148 (paperback) | ISBN 9781725323179 (library binding) | ISBN 9781725323155 (6 pack)
Subjects: LCSH: Plastics--Environmental aspects--Juvenile literature. | Plastic scrap--Environmental aspects--Juvenile literature. | Plastics--History--Juvenile literature.
Classification: LCC TD798 .F87 2022 | DDC 363.738--dc23
LC record available at https://lccn.loc.gov/2019052501

Manufactured in the United States of America

CONTENTS

THE PROBLEM WITH PLASTIC 4
THE BENEFITS OF PLASTIC 8
ENVIRONMENTAL WAKE-UP CALL 10
PLASTIC POLLUTION 14
MYTHS AND FACTS 18
TAKE ACTION! 20
NEW SOLUTIONS 26
HOW CAN YOU HELP? 28
GLOSSARY 31
INDEX 32
PRIMARY SOURCE LIST 32
WEBSITES 32

CHAPTER ONE

THE PROBLEM WITH PLASTIC

If you look around at cars, toys, and many other things, you may notice a very common material used in all of them. That substance is called plastic, and it's very useful. Unfortunately, careless and wasteful practices with plastic have caused many problems, not only for people but also for countless species of plants and animals and the entire planet.

Plastic doesn't naturally occur on Earth. It's manufactured from fossil fuels such as oil, natural gas, and coal. Plastic hasn't been around for a very long time. People first created it during the late 19th century, a little more than 100 years ago. In that short span of time, we've managed to create 9.2 billion tons (8.3 mt) of plastic. Of that, over 6.9 billion tons (6.3 mt) has become waste on land and in the oceans. Scientists have even found small particles of plastics, called microplastics, in very remote places. In a recent study, a 2.1-pint (1 L) sample of sea ice from the Arctic Ocean was found to contain 14,000 particles of plastic.

There have been many efforts to recycle discarded plastics so they can be made into something new. Unfortunately, recycled plastics aren't very common. According to research from the University of Georgia, over 90 percent of plastics have never been recycled.

It's obvious the plastic problem is a big one, but it's not being ignored. Efforts by governments all around the word, as well as those by people just like you, are working to break our wasteful habits with plastics.

Plastic pollution is a worldwide problem. Over 300 million tons (272.2 million mt) of new plastic is produced every year. About 8 million tons (7.3 million mt) of plastic winds up entering the world's oceans.

Before plastics were invented, other raw materials such as wood, metal, cotton, glass, and animal bones were used to manufacture things. In the 1800s, people used elephant tusks to make many items, including combs, piano keys, and balls for billiards, a game similar to pool. The use of elephant ivory caused a sharp decline in the elephant population. In 1867 the *New York Times* warned that elephants were in danger of being "numbered with extinct species."

People started looking for substitutes for ivory and other materials, such as turtle shells. In 1869, a chemist named John Wesley Hyatt invented a new material from cotton cellulose, a plant polymer. A polymer is a large molecule that's made of many smaller, repeating molecules. Hyatt and his brother marketed the new material and called it celluloid. Although it was too hard to be used for billiard balls, it was good for making many other things, including combs, buttons, toys, and dice.

LEO BAEKELAND

Bakelite products such as these art deco radios and clocks are now considered collectible.

Leo Baekeland created the first synthetic plastic in 1907. "Synthetic" means it's made entirely from compounds not found in nature. Using various byproducts from coal use, Baekeland created a plastic that could be molded into different shapes. The plastic was called Bakelite. It was used to make radios, phones, car parts, and even jewelry, which is now very collectible.

Soon, people created many other plastic formulas, all with names as unique as their properties. They included polystyrene, Styrofoam, and nylon. In time, manufacturers used nylon to make women's stockings, parachutes, rope, and toothbrush bristles.

CHAPTER TWO

THE BENEFITS OF PLASTIC

During World War II (1939 to 1945), plastic production in the United States increased by 300 percent. Many traditional materials were expensive or hard to come by during this time, and plastics helped fill in the gap. From parachute cords and ropes made of nylon to aircraft windows made out of plexiglass, plastic was becoming more common and more useful.

After World War II, the plastic industry began manufacturing products for buyers as inexpensive alternatives to traditional household items. Tupperware, the now-common food storage container, was introduced in 1948. In the 1950s, a hard plastic called Formica was used to make colorful kitchen countertops and tables.

A Swiss engineer named George de Mestral invented the "hook-and-loop" fastener Velcro in 1941. Mestral was inspired by natural cockleburs he found stuck to his pants in the woods. Today Velcro is used for many things. It has replaced safety pins on diapers and shoelaces on kids' sneakers.

VELCRO

Plastics have also made vehicles safer. Because plastic is lightweight and still very strong, plastic parts have replaced many heavier metal parts in cars and airplanes. Safety airbags in cars are made from nylon.

Plastics are also used in the healthcare industry. Many items, such as surgical gowns, face masks, gloves, and IV bags, are all made from plastics.

A durable plastic called Kevlar is used in bulletproof vests to protect police. There's no doubt that the benefits and uses of plastics today are numerous.

In this picture, taken in 1946, women wait in line to buy nylon stockings. During World War II, nylon was used to make war materials, creating stocking shortages throughout the United States.

CHAPTER THREE

ENVIRONMENTAL WAKE-UP CALL

During the 1960s, the average person in the United States used about 30 pounds (13.6 kg) of plastic yearly. Now the average person uses 220 pounds (99.8 kg) of plastic yearly. As of 2019, global plastic production is at 300 million tons (272.2 million mt) a year. If the present trend continues, there could be 12 billion tons (10.9 billion mt) of plastic in landfills by 2050.

By the 1960s, people were beginning to take notice of plastic waste. The first record of a plastic bag in the ocean took place in 1965. Scientists using plankton-sampling equipment accidentally snagged a plastic bag.

Other environmental incidents drew people's attention. Smog, which is heavy ground-level air pollution, was noticed in large cities such as New York and Los Angeles in the late 1950s and early 1960s. Air quality and pollution became a topic of concern.

In 1969, an oil platform spilled 100,000 barrels of oil into the Pacific Ocean off the coast of Santa Barbara, California. The spill washed up along 35 miles (56.3 km) of coastline and killed many animals. Public outcry resulted in many new environmental laws and regulations, including a ban on offshore oil-drilling platforms.

The modern environmental movement was taking shape. By 1970, something special took place. People held the first Earth Day demonstrations across the United States on April 22, 1970. Many consider the first Earth Day to be the official start of the modern environmental movement.

Yukihisa Isobe designed the first Earth Day poster, pictured here, for the Environmental Action Coalition.

As plastic use in America increased, so did litter awareness. In 1953, a group of leaders formed a nonprofit anti-litter organization called Keep America Beautiful. The organization launched public service announcements to promote responsible anti-litter habits.

To help control air pollution, Congress passed the Clean Air Act in 1963. By 1965, the law was amended to include the Solid Waste Disposal Act. This was the first federal law that regulated the disposal of waste with the future of the environment in mind.

Plastic waste that didn't make its way into garbage dumps was more than an eyesore. It affected animals badly. In 1969, biologist Eugene Kridler and zoologist Karl Kenyon were the first to report that Hawaiian seabirds had mistakenly eaten plastic.

Increased efforts to take care of the environment continued. In 1970, President Richard Nixon signed an executive order and established the Environmental Protection Agency (EPA).

Throwing plastic into the garbage didn't help reduce the amount that was being produced. To reduce plastic production, people needed to embrace recycling. In 1971, Oregon passed the first bottle deposit law. In 1980, Woodbury, New Jersey, became the first city in the United States to start a recycling program.

PRESIDENT RICHARD NIXON

In the 1970s, there were many anti-litter ads throughout the United States. This poster shows many common items that can be recycled rather than being discarded.

Unfortunately, early efforts to recycle plastic didn't go far enough. Plastic items were inexpensive to produce, and in 1979, the plastic bag was introduced in grocery stores across the United States. Despite efforts, the plastic problem was getting worse. Today, people use 10 billion plastic bags in the United States every year.

CHAPTER FOUR

PLASTIC POLLUTION

About 71 percent of Earth's surface is covered by water. A great deal of plastic debris enters the oceans each year. The estimates vary, but the numbers are big. Somewhere between 8.8 million and 20 million tons (8 million and 18.1 million mt) of plastics enter the oceans every year. About 40 percent of all plastic waste comes from single-use plastic items. These are items that are only used once before being thrown away. Some examples of single-use items are plastic beverage bottles, food packaging, straws, and bags. Other plastic items that wind up in the oceans come from discarded fishing gear, such as nets and fishing lines.

Wind and ocean currents around the world, as well as Earth's rotation, create large circular ocean currents called gyres. There are five major ocean gyres around the world. Because of their circular currents, great amounts of plastic debris accumulate in these gyres.

In 1997, a sailor named Charles Moore discovered plastic debris floating everywhere in the North Pacific Ocean gyre. Today, that gyre has become known as the Great Pacific Garbage Patch. Moore was so upset by the garbage he found floating in the gyre that he dedicated his life to studying plastic waste in the oceans.

Today, there's about 87,000 tons (79,000 mt) of plastic in the Great Pacific Garbage Patch, most of which comes from discarded fishing gear. People are paying more attention to garbage patch gyres, so efforts to clean up plastics from the world's oceans have increased.

Plastic litter kills as many as 100 million ocean animals every year. Leatherback sea turtles often mistake plastic bags for their natural food, jellyfish, which results in 100,000 deaths each year.

Plastic ocean garbage gets a great deal of attention. But there's growing evidence that land-based plastic pollution is just as bad, if not worse. Just as sea animals consume plastic, land animals do too. In some cases, land animals have gotten caught in plastic twine, and some have had their heads stuck inside plastic food containers. Birds will often use plastic debris to make nests, which may be harmful to their young.

Because plastic is so durable, it breaks apart into smaller pieces long before it biodegrades into less harmful parts. No one knows exactly how long it takes plastic to completely biodegrade, but some estimate it may take as many as 450 years. Some think that plastic may never completely biodegrade.

Small pieces of plastic are called microparticles. These are smaller than 0.2 inch (5 mm). Nanoparticles are much smaller. These plastic particles get into fresh water, soil, and sewage, or waste. Over time, they enter the food chain and wind up inside animals and humans. These plastic particles pose a risk to both plants and animals. Smaller pieces of plastic release harmful chemicals more easily than larger ones do. These chemicals can pollute the water or soil. These plastics and chemicals can hurt animals that ingest, or consume, them. Larger plastic pieces can block animals' intestinal tracts, sometimes leading to death.

In order to solve the plastic waste problem, people are now examining single-use items more than ever before.

Plastic land pollution poses health risks to plants and animals, including humans. More than half of all plastics ever produced sit in landfills where harmful chemicals can leach, or leak, out and contaminate soil and groundwater.

CHAPTER FIVE

MYTHS AND FACTS

There are many myths and misunderstandings about plastic. The wrong information can lead to poor decisions about the best way to be responsible with this material. Here are a few common myths and facts about this useful polymer, which has also become a very large problem.

Myth: Plastic eventually biodegrades and will be harmless to the environment.

Fact: Because plastic hasn't been on Earth for very long, no one is exactly sure how long it takes to break down. Some of the first plastics created are still in the environment.

Myth: In the world's oceans, there are gigantic floating islands of plastic garbage that you can walk on.

Fact: Despite the many tales about the Great Pacific Garbage Patch and other gyres of plastic in the world's oceans, they aren't giant floating islands. None of the concentrations of plastic in the oceans can be walked on. Most of the plastic contained in the sea is under the surface and can't be seen.

Myth: Not all plastic is recyclable.

Fact: Most plastic can be recycled, but over 90 percent of all plastics that have been produced have never been recycled. If recycling is available where you live, check to see what kinds of plastics are accepted for recycling. Recycling sorting centers are constantly improving and accepting more types of plastics. Even if a certain recycling center doesn't take a certain plastic, that doesn't necessarily mean it can't be recycled.

Not all plastic pollution is visible. The Great Pacific Garbage Patch is partly made up of huge amounts of microplastics, which make the water look cloudy.

CHAPTER SIX

TAKE ACTION!

Using a single plastic straw or plastic bag may not seem like a big deal, but personal plastic usage adds up over time. Despite efforts, plastic pollution continues to increase. Since 2000, humans have used more plastic than they'd used in all the years before. In 2017, the average person from North America used 220 pounds (99.8 kg) of plastic. Most of this was from single-use plastics. If current usage rates continue, that number will double by 2030.

In 2012, Hawaii became the first state to ban plastic bags with local laws. In 2014, California became the first state to ban single-use plastic bags with a state law. As of 2019, Connecticut, Delaware, Maine, New York, Oregon, and Vermont have also banned single-use plastic bags.

Some of these laws don't go into effect right away. But there's no need to wait for a new law to reduce personal plastic usage. Here are a few steps everyone can take now to cut down on plastics and make a difference:

- Stop using plastic straws. You could use a reusable straw instead.
- Stop using disposable cups when you buy ice cream. You can eat a cone and it doesn't require a spoon.
- Quit using disposable lunch bags and plastic wrap. Instead, make a reusable cloth lunch bag. You can also now find reusable food wrappers.
- Use reusable water bottles instead of single-use bottles.

Using plastic alternatives can have a big impact. This way, everyone can help with the global plastic cleanup.

Make a difference! If a person uses reusable shopping bags for their entire life, they could save as many as 22,000 plastic bags from being discarded.

For many decades, China imported about half of all worldwide plastics collected for recycling. In 2017, in an effort to improve the environment there, China stopped accepting plastic from around the world.

The sudden ban created an oversupply of plastic waste and left many countries wondering what to do with their plastics. This also acted as a wake-up call that moved some countries to action. In 2018, the European Parliament voted to ban single-use plastics across the entire continent. The law was approved in 2019 and will take effect in 2021. There are also efforts to recycle 90 percent of all plastic bottles in Europe by 2029.

The global adjustments haven't been easy. As a result of China's decision, many cities in the United States have recyclable plastics piling up. Several cities have stopped recycling altogether.

Balinese sisters Isabel and Melati Wijsen said their efforts were inspired by the work of leaders such as Nelson Mandela and Mahatma Gandhi.

Long before China's nationwide ban on imported plastic waste, two sisters from Bali, Indonesia, were already working towards solutions. Isabel and Melati Wijsen were only 10 and 12 years old in 2013, when they got fed up with the plastic waste in their country and decided to fix the problem. Indonesia is the second-largest plastic polluter of the world's oceans.

With signature campaigns and cleanup efforts, the determined duo got the attention of the Bali government, which is now working to reduce plastic bag use through awareness and bans. The sisters also started Bye Bye Plastic Bags, an international nongovernmental youth organization helping reduce plastic waste in over 25 locations.

In 2011, a Dutch teenager named Boyan Slat was diving off the coast of Greece when he saw more plastic bags in the sea than fish. Overwhelmed by what he witnessed, Slat decided to take action. In high school and college, Slat continued to search for a way to rid the world's oceans of plastic waste.

Many of the solutions proposed required ships dragging nets that would also wind up capturing sea animals. Slat thought of an anchored invention that would float and use ocean currents to collect plastic debris while keeping animals safe. The floating collection system would then carry the captured debris onto a platform for removal.

After crowdsourced fundraising, Slat founded The Ocean Cleanup in 2013. According to its website, theoceancleanup.com, "The Ocean Cleanup, a nonprofit organization, is developing advanced technologies to rid the world's oceans of plastic."

Slat sent the first cleanup system off the coast of Japan in 2016, but it didn't work as planned. By 2019, after some trial and error, The Ocean Cleanup had a new system that efficiently captured ocean plastics. Slat and his team of more than 80 engineers and scientists hope to scale up their efforts and eventually send out a larger fleet. The Ocean Cleanup estimates that with reduction of plastics entering the oceans and cleanup systems in all the world's gyres, 90 percent of plastic waste in the world's oceans could be removed by 2040.

In addition to his system for removing plastic from the world's oceans, Boyan Slat has also created an anchored river barge called the Interceptor. It can catch as much as 55 tons (50 mt) of garbage per day, preventing the trash from entering oceans.

CHAPTER SEVEN

NEW SOLUTIONS

The world is too dependent on plastics. Their durable and malleable properties allow for many different uses. However, their strengths are also why they don't break down when discarded and have been increasingly polluting the planet since the 1950s.

Because the problem is increasing and the main culprits are single-use plastic items, many people are working hard to find plastic alternatives that could be easier on animals and the environment where they often wind up.

One possible solution is a new biodegradable plastic manufactured from sugar and carbon dioxide. Researchers at the University of Bath developed this new form of plastic, which can break down into its original components with enzymes found in soil bacteria.

Disposable water bottles create a lot of waste. According to the Container Recycling Institute, in 2014 there were roughly 300 plastic bottles manufactured for every person living in the United States. However, a sustainable packaging start-up has used seaweed and other plants to make a new material called Notpla. The material is also the company's name. Because Notpla comes from organic material, it takes only a few weeks to biodegrade. The company Notpla also manufactures small, clear, edible bubbles that can store water or other liquids. This material is called Ooho. The company has also created a Notpla liner for cardboard food containers that are normally coated with plastic. This new material might just be a solution to the plastic bottle problem.

Unlike plastic bottles, you can eat the material that holds this drink!

CHAPTER EIGHT

HOW CAN YOU HELP?

Everyone can help solve the global plastic problem. Here are some ideas to get you and your friends started on removing plastic waste from the environment:

- Start a plastic bottle collection group. Many states, including New York, California, Maine, Hawaii, and Vermont, have deposit laws so any plastic bottles collected there can be redeemed for some money. You'll be helping clean up where you live and earning some money in the process!
- Plastic bottles can be used to create new things. Using discarded items and repurposing them into something useful is called upcycling. In 2016, Daniel Edwards started a Kickstarter project to create a 3-D pen called Renegade. This pen uses plastic bottles for its source material instead of costly new materials. Some other 3-D pen companies are catching on. Nefilate and Refil both use recycled plastic for their 3-D pens. With so much plastic lying around, why not recycle it yourself to make something new?
- Upcycling plastics is a great way to get creative. Start a project creating art, planters, magazine racks, or baskets. They can all be made from discarded plastic. Remember to collect and use only discarded plastics rather than buying anything new. The goal is to clean up.
- Start a home recycling station. Recycling items often need to be sorted before they can be discarded. Use some cardboard boxes that were headed to the trash as containers for metal, plastic, and glass. Label the boxes so everyone knows exactly what goes where.

This greenhouse was made from wood and discarded plastic bottles. What will you make with the plastic you find?

Plastic pollution can come from many unlikely sources but so can solutions. In 1997, a cargo container ship lost many containers filed with LEGO toys near Cornwall, England. LEGO pieces are still washing up in beaches along the coasts of Cornwall today. In 2018, LEGO released sustainable, plant-based plastic bricks. That same year, the company recycled 93 percent of all company waste. It also pledged to make all bricks from sustainable sources by 2030. The new bricks are easy to recycle but will not biodegrade.

Traditional plastic recycling can be easily ruined if items are dirty, contaminated, or mixed with other materials. However, a company called Agilyx is using a new cutting-edge form of chemical recycling. Unlike traditional recycling, which reuses plastics, Agilyx uses chemicals to break plastic down into its original molecular components. Agilyx even accepts single-use items that are rarely recycled, such as those made of Styrofoam.

Because plastics are durable and last for such a long time, MacRebur, a company from Scotland, is taking nonrecyclable plastics and repurposing them into durable roads. MacRebur has used its new plastic pavement mixture to build roads all around the world, including a highway in England and a road in San Diego, California.

People are still coming up with unique ideas of how to deal with the plastic problem. Everyone will need to contribute with help, ideas, solutions, and hard work. Solutions can come from anyone, even you. No idea is too small, and every effort helps.

GLOSSARY

accumulate (uh-KYOO-myuh-layt) To gather or build up.

biodegrade (BY-oh-dih-grayd) To naturally break down.

celluloid (SEL-yuh-loid) A plastic made mainly from plant-based materials.

contaminate (kuhn-TAA-muh-nayt) To pollute.

crowdsource (KROWD-sohrs) To gather services, ideas, or funds from a large group of like-minded people, typically from the internet.

culprit (KUHL-pruht) Something to blame for something.

discard (DIS-kard) To throw away.

Earth Day (URTH DAY) April 22, a day set aside around the world to help promote environmental awareness for planet Earth.

edible (EH-duh-buhl) Safe to eat.

enzyme (EHN-zym) A chemical substance that helps natural processes in plants and animals.

fossil fuel (FAH-suhl FYOOL) A fuel—such as coal, oil, or natural gas—that is formed in the earth from dead plants or animals.

Great Pacific Garbage Patch (GRAYT puh-SIH-fik GAHR-bij PACH) A region of circular currents located in the northern Pacific Ocean that captures ocean waste.

import (IM-pohrt) To bring in from another country.

Kevlar (KEHV-lahr) A brand of man-made fabric that is very strong and thick.

malleable (MA-lee-uh-buhl) Able to be stretched or bent into various shapes with breaking.

parliament (PAHR-luh-muhnt) A lawmaking body.

plankton (PLANK-tuhn) A tiny plant or animal that floats in the ocean.

sustainable (suh-STAY-nuh-buhl) Able to last a long time.

technology (tek-NAH-luh-jee) A method that uses science to solve problems and the tools used to solve those problems.

INDEX

A
Agilyx, 30

B
Baekeland, Leo, 6, 7
Bye Bye Plastic Bags, 23

C
celluloid, 6
China, 22, 23
Clean Air Act of 1963, 12

E
Earth Day, 10, 11
Environmental Protection Agency (EPA), 12

G
Great Pacific Garbage Patch, 14, 18, 19

H
Hyatt, John Wesley, 6

I
Indonesia, 23

K
Keep America Beautiful, 12
Kenyon, Karl, 12
Kevlar, 9
Kridler, Eugene, 12

L
LEGO, 30

M
MacRebur, 30
Mestral, George de, 8
microplastics, 4, 19
Moore, Charles, 14

N
Nixon, Richard, 12
Notpla, 26

O
Ocean Cleanup, 25

R
recycling, 12, 18, 22, 26, 28, 30

S
Slat, Boyan, 24, 25
Solid Waste Disposal Act, 12

W
Wijsen, Isabel and Melati, 23
World War II, 8, 9

PRIMARY SOURCE LIST

Page 9
Women standing in line for nylon stockings at Miller's Department Store. Photograph. Ed Westcott. Oak Ridge, Tennessee. January 4, 1946.

Page 11
Earth Day poster. Designed by Yukihisa Isobe. 1970. Now kept at the Library of Congress.

Page 13
Recycling poster. 1970. Now kept at the Library of Congress.

WEBSITES

Due to the changing nature of Internet links, Rosen Publishing has developed an online list of websites related to the subject of this book. This site is updated regularly. Please use this link to access the list: www.powerkidslinks.com/SOGI/earthsplastic